U0013790

Life Lessons
I Learned
from My Cat

Illustrated by Jamie Shelman

走路要有喵態度

沙發上的心靈大師，
給人類的100則貓生哲學

潔咪·薛曼 ——— 繪　　陳采瑛 ——— 譯

 擊掌推薦

　　動物是人類最好的導師，此話所言不假，從貓咪身上可以學到的東西可多了！插畫家潔咪‧薛曼透過觀察鄰居的貓咪獲得靈感，她筆下的貓咪們身材圓潤、四肢細小，配上詼諧的神韻和動作，外型十分討喜，加上精闢的智慧小語，形成一幅幅令人莞爾的小插畫，讓人忘卻現實中的煩惱。

—— Lynol 圖文作家

　　當我收到《走路要有喵態度》的推薦序邀約時，我立刻打開這本書，邊讀邊望著我身旁那隻躺在窗邊的貓，每翻一頁就會心一笑。心裡想著：是啊，這些重要的人生大道理，在貓身上看起來似乎如此簡單——需要休息時就睡覺、捍衛自己所要的、不在乎他人眼光、不斷嘗試、不害怕表達你的喜愛、做你自己。我的貓每天都教我原來生活可以這樣過，也希望這本書可以帶著你重新體會生命。

—— 留佩萱 美國心理諮商博士、執業諮商師

翻開這本書,在你的心裡豢養一隻貓吧!

累了就休息、想睡就打盹、做不好就把紙撕爛,打從內心成為獨立又特別的自己,不需要擔心別人的眼光⋯⋯這些勵志書上的句子你都懂,但你都達成不了;而如果你心裡有一幅貓咪的圖像,想想「如果是貓會怎麼做」,可能一直困住你的那些瓶頸,就會豁然開朗。

——海苔熊 心理學作家

一本適合放在手邊,隨時隨地拿起來看都能鬆一口氣微笑的生活祕笈(前提是你的貓咪沒有蹲在上面)。

——路嘉欣 副業歌手、演員,主業貓奴

夜深了，如果你還在煩惱明天的簡報，本書的喵喵主角會大聲告訴你：「解決問題的手段，就是去睡個覺！」拜讀這本充滿深度、廣度、靈活度的貓式哲學，瞬間被貓掌加持，醍醐灌頂，人生從此大智如貓、廢柴如貓、快活如貓。

　　　　　　　　　　　　　　　　　——貓小姐 文字工作者

　　欣賞貓，始於顏值，敬於優雅，合於性格，久於互敬，忠於自己。

　　有貓人生就是這樣，一切得都從頭學起。課本上沒教過的道理、美術課沒畫過的觸筆、還給老師已生疏的英語，因為這本書，可以再次找回快樂的自己。

　　　　　　　　　　　　　　　　　——貓夫人 攝影工作者

感到生活壓力如龐然大物，拔山倒樹而來，悲憤當人不如當塊叉燒嗎？

等等！別急著抹了調味料便跳進烤箱裡，不妨學學喵星人悠然自得之生活哲學。這本書以柔和色彩、詼諧畫風，一筆筆勾勒出貓咪們可愛靈動神情。主子頒布一百條聖旨，掌把掌地教你人生要有喵態度，才能揮別愁雲慘霧！既然領旨了還不快快謝恩，莫忘時時勤拂拭，莫使惹塵埃啊！

——螺螄拜恩 暢銷作家

本書獻給我鄰居的貓，布魯克西，
因為牠每天早上都在窗前等我們

我這輩子很愛貓，也和很多隻貓一起生活過，我知道透過觀察貓科朋友能夠學到許多人生智慧。從如何生活、如何去愛、如何得到你想要的，到如何與自己和平共處，以及明白生命中最重要的事情……也就是食物，睡眠和一點友情！

　　在這個快步調、使人分心的世界，我發現自己現在更想從貓（基本上是從我鄰居的貓，布魯克西）身上尋求安慰、靜心、喜悅與小小的開示。貓可以既有智慧又傻里傻氣、黏人又冷眼旁觀、簡單又獨一無二，閒閒沒事做而樂在其中，這牠們超會的！我希望你透過這本書能了解，原來可以從自家的貓身上習得一些度人生的道理。然後有一天我們都能學會只是靜靜坐著，歡喜地瞇著眼，心滿意足直到下一餐開飯！

潔咪

要固定伸展

Stretch regularly

充分休息

Get plenty of rest

……因為打盹
沒什麼好丟臉的

...Because naps are never to be ashamed of

保持溫暖

Stay warm

享受你的安靜時刻

Enjoy your quiet time

維持體面的儀容

Maintain a well-groomed appearance

多吃些魚

Eat more fish

沐浴在
陽光下

Soak up the sun

只吃你要吃的

Only eat what you want

有機會就梳洗

Wash when you can

認真打理才有美麗的毛皮

A beautiful mane takes serious work

……完美的指甲也是

...And so does the perfect set of nails

當個晨型人

Be an early riser

……但是可以在床上待久一點

...But it's okay to stay in bed a little bit longer

作白日夢

Daydream

用你的話語表達主張

Speak up and use your voice

儲備精力

Conserve your energy

跳脫
框架思考

Think outside the box

要有耐心⋯⋯

Be patient...

……緊盯著目標

...And keep your eye on the prize

學習溝通的藝術

Learn the art of negotiation

就是要直接

Be direct

給予正面的回應

Give positive feedback

用眼神交流

Make eye contact

碎掉所有文件

Shred all documents

別灰心

Don't be discouraged

不要
工作過度

Don't work too hard

要對房裡那一個
不喜歡你的人特別殷勤

Be especially attentive to the one person
in the room who doesn't like you

對自己的成就感到滿意，
即使很小

Be pleased with your achievements,
however small

把眼裡的殺氣練到呼嚕完美

Purrfect your death stare

把撲克臉練到呼嚕完美

Purrfect your poker face

打響
你的
名聲

Make your mark

走路要有喵態度

Walk with (c)attitude

……爪子要保持銳利

...And keep the claws sharp

保持專注

Stay focused

一旦下定決心，
就不動搖

Once you've made up your mind, don't change it

目標要高遠

Aim high

解決問題的最佳手段
就是睡個覺

The best solution to a problem is a nap

焦慮
沒什麼
好可恥

Anxiety is nothing to be ashamed of

活在當下

Live in the moment

要有適應力

Be resilient

不要為小事煩心

Don't sweat the small stuff

討厭社交沒關係

Being anti-social is okay

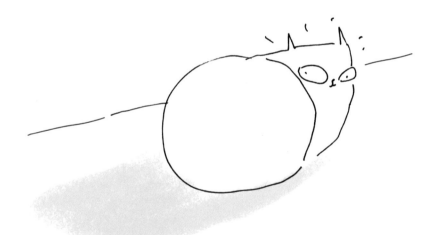

花時間自我反省

Take time to reflect

對抗你的恐懼

Confront your fears

慢下來

Slow down

只在覺得有勁的時候運動

Exercise only when you feel like it

到
舒適圈外
生活

Live outside of your comfort zone

對事不對人

Nothing is personal

莫名的害怕是正常的

Irrational fears are normal

將你所擁有的發揮到極致

Make the most of what you've got

喝水有益心靈

Drinking water is good for the soul

先照顧好自己

Look after yourself first

別怕讓人知道你喜歡他們

Don't be afraid to let someone know you like them

當他們進門時，永遠要看起來很開心
Always look pleased when they come through the door

做你自己，
那個對的人
會因此愛你

Be who you are and the right person
will love you for it

善於接受情意

Be good at receiving affection

但不可索求無度

But don't be needy

和你所愛的人無論何時
都保持至少三尺遠

Stay at least ten feet away from
your loved one at all times

女人和貓想做什麼就做什麼

Women and cats will do as they please

……而男人與狗應該放輕鬆
並且習以為常

...And men and dogs should relax and get used to it

先愛你自己

Love yourself first

照你的意願接納愛

Accept love on your own terms

對不崇拜你的人視而不見

Ignore anyone who doesn't worship you

縱身一躍前先看清楚

Look before you leap

絕對
不要信任
不喜歡貓
的男人

Never trust a man who doesn't like cats

相信你的直覺

Trust your intuition

要好奇

Be curious

忠於自我

Be yourself

要獨立

Be independent

讓人難以捉摸

Be elusive

有時你一躍起就失足……

Sometimes you will leap and fall...

……但你可以逢凶化吉

...But you can land on your feet

踩線看看

Test the boundaries

隨時保持你的尊嚴

Preserve your dignity at all times

做一個優秀的傾聽者

Be a good listener

把你的九條命發揮到極致

Make the most of your nine lives

探索世界

Explore the world

走到哪睡到哪

Nap anywhere

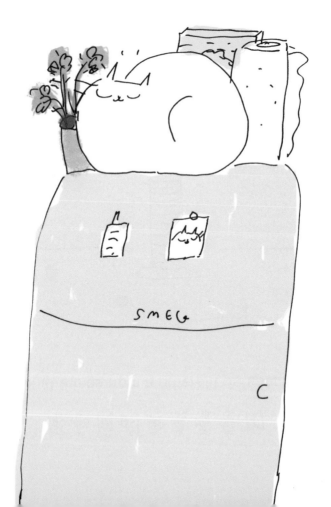

Look at things from a different perspective

用火同的角度看事物

不用擔心別人怎麼看你

Don't worry what others think of you

一聲哈氣勝過千言萬語

A hiss is worth a thousand words

要易於取悅

Be easily entertained

用賣萌
來避開
殺身之禍

Get away with murder by looking cute

不計代價都要成為注目焦點

Be the centre of attention at whatever cost

一本好書勝過所有

There's nothing better than a good book

學著無論在何處
都能泰然自適

Learn to make yourself
comfortable wherever you are

永遠不要失去玩心

Never lost your playfulness

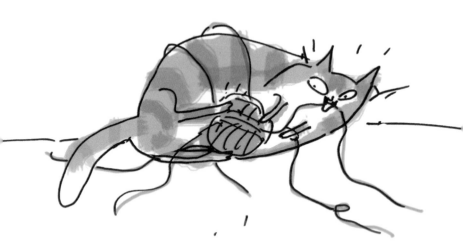

堅守立場不退讓
Stand your ground

要他人尊重
你的意願

**Be taken on
your own terms**

總是慷慨給予

Always give generously

慎選朋友

Choose your friends wisely

要經常要笨

Be silly...and often

親近你的敵人

Keep your enemies close

擁抱
你的特色

Embrace your weird

對小孩有耐心

Be tolerant of children

讚揚一家子的用餐時光

Celebrate family mealtimes

如果你想吸引誰的注意力，

If you want someone's attention,

就坐到電視前面

Sit in front of the TV

永遠保衛你的家

Always defend your home

屋內亂七八糟才是快樂的家

A chaotic home is a happy home

別怕老鼠

Don't be afraid of mice

怎麼穿，
我說了算

Never let anyone dress you

屋裡最棒的位置
就是有人正坐著的地方

The best seat in the house
is the one someone is already sitting on

| 繪者簡介 |

潔咪・薛曼（Jamie Shelman）

畢業於羅德島設計學院（RISD），以藝術創作為業，同時也經營文創品牌「The Dancing Cat」，商品特色是獨特的貓設計，鄰居的貓布魯克西是她的工作室常客兼創作繆斯。目前居於美國巴爾的摩。

| 譯者簡介 |

陳采瑛

畢業於中央大學英文所，偶爾客串當譯者。譯過許多以動物為主角的繪本，包含《一顆海龜蛋的神奇旅程》、《頑固的鱷魚奶奶》、《燕子的旅行》、《歡迎光臨蟲蟲旅館》、《鳥巢大追蹤》等。

走路要有喵態度

沙發上的心靈大師，給人類的100則貓生哲學

圖──潔咪·薛曼（Jamie Shelman）
文──Michael O'Mara Books Ltd
譯──陳采瑛

責任編輯──陳嬿守
副 主 編──陳懿文
封面設計──謝佳穎
內頁排版──陳春惠
行銷企劃──鍾曼靈
出版一部總編輯暨總監──王明雪

發行人──王榮文
出版發行──遠流出版事業股份有限公司　104005 台北市中山北路一段11號13樓
　　　　　　電話／(02)2571-0297　傳真／(02)2571-0197　郵撥／0189456-1
著作權顧問──蕭雄淋律師
□2020年 2 月 1 日 初版一刷
□2023年 1 月20日 初版四刷

定價──新臺幣320元（缺頁或破損的書，請寄回更換）
有著作權·侵害必究　Printed in Taiwan
ISBN──978-957-32-8698-1

ylb 遠流博識網 http://www.ylib.com　E-mail: ylib@ylib.com
遠流粉絲團 https://www.facebook.com/ylibfans

....

國家圖書館出版品預行編目 (CIP) 資料

走路要有喵態度：沙發上的心靈大師，給人類的100則貓生哲學
　/ 潔咪‧薛曼（Jamie Shelman）繪；陳采瑛譯. -- 初版. -- 臺北
　市：遠流, 2020.02
　　　面；　公分
　譯自 :Life lessons I learned from my cat
　ISBN 978-957-32-8698-1（精裝）

1.貓　2.文集　3.生活指導

437.3607　　　　　　　　　　　　　　　108021886